Five-Fold Ministry

Christ's Gift to the Church

By Bonnie Sherwood

MACEDONIAN PRESS

First Edition

ISBN: 978-1-968118-04-4

Published by Macedonian Press, Bowling Green, KY

Some scriptures are Bonnie Sherwood's personal paraphrase (TBV), TheBonVersion

Please note that our publishing style capitalizes certain pronouns in scripture that refer to the Father, Son, and Holy Spirit.

This book is a work of testimony, teaching, and spiritual encouragement. It is not intended to replace professional medical or psychological advice. Readers are encouraged to seek the Lord and consult trusted advisors when needed.

Table of Contents

Introduction

The Eternal Purpose of the Five-Fold Ministry

"And He Himself gave some to be apostles, some prophets, some evangelists, and some pastors and teachers, 12for the equipping of the saints for the work of ministry, for the edifying of the body of Christ, 13till we all come to the unity of the faith and of the knowledge of the Son of God, to a perfect man, to the measure of the stature of the fullness of Christ;"

~Ephesians 4:11–13 (NKJV)

God's Wonderful Plan

Have you ever stopped to marvel at how intentional God is? He never leaves His people without guidance, without nurture, without a living witness of His presence. From Eden's garden, to the wilderness tabernacle, to the temple in Jerusalem, His heart has always been to dwell among His people and shepherd them into maturity. The Five-Fold Ministry is a continuation of this story. It is heaven's design for the Church in every generation; a reflection of Christ Himself distributed among His body.

When Paul revealed these gifts to the Ephesian church, he was not proposing a new church model or drafting a system of hierarchy. No—he was pulling back the curtain on eternity. He wanted believers to understand that these gifts are not man-made positions but divine impartations. The apostle, prophet, evangelist, pastor, and teacher are not separate entities fighting for prominence. They are expressions of the one Lord Jesus, each carrying a unique aspect of His nature, each bearing a part of His heart.

Think of a diamond held up to the light. Each facet catches the sun in a different way, yet together they reflect the brilliance of the whole. That is what the Five-Fold Ministry does: it refracts the light of Christ so that His body can see Him more clearly. Apostles reveal His authority and pioneering Spirit. Prophets echo His voice and unveil His heart. Evangelists embody His relentless pursuit of the lost. Pastors mirror His compassion and care. Teachers unfold His wisdom and truth. None of these gifts alone display His fullness, but together, they make Him known.

Unity in Diversity

We live in a fractured age. Ministries compete for recognition, leaders burn out from carrying weights they were never meant to bear, and the body of Christ sometimes limps along because it leans too heavily on one gift while neglecting the others. Have you ever seen a hand try to function with only two fingers? It can still grip, but it is awkward, incomplete. So it is with the Church. The five gifts must move together—like the fingers of one hand, guided by the Head, Christ Jesus.

When The Five flow in unity, the Church becomes balanced, healthy, and fruitful. The apostle sets vision, the prophet guards purity, the evangelist ignites passion, the pastor nurtures wholeness, and the teacher anchors truth. Remove one, and the body falters. Elevate one above the rest, and imbalance sets in. But when each part honors the others, when the hand opens wide in obedience to Christ, the Church begins to look like Jesus in the world.

Equipped for Eternity

Notice Paul's language: these gifts were given *"until we all attain to the unity of the faith... to the measure of the stature of the fullness of Christ."* The purpose is eternal. This is not about building earthly empires or creating polished programs. The goal is to prepare the Bride for her Bridegroom, to shape a people who are mature, steadfast, and radiant with His presence.

Beloved, the Five-Fold Ministry is not ultimately about leadership structures—it is about love. It is Christ's love poured out through His people, so that no one lacks what they need to grow. He distributes Himself across His body, ensuring that every believer, every local church, every community of faith, has access to His guidance, His care, His teaching, His truth, His zeal.

And here is the eternal perspective: these gifts will not be needed forever. One day, when we see Him face to face, prophecy will cease, teaching will be swallowed up in perfect knowledge, shepherding will be fulfilled in the Great Shepherd's embrace. But until that day, He has given us these treasures to mature us and keep us pressing toward the fullness of Christ making love our motivation in all things.

"Love never fails. But whether there are prophecies, they will fail; whether there are tongues, they will cease; whether there is knowledge, it will vanish away. 9For we know in part and we prophesy in part. 10But when that which is perfect has come, then that which is in part will be done away."
~1 Corinthians 13:8-10 (NKJV)

An Invitation to the Journey

In the pages that follow, we will walk slowly through each of these five gifts. We will listen for the heartbeat of the apostle, the prophet, the evangelist, the pastor, and the teacher. We will reflect on how each one expresses the character of Jesus, and we will consider how they all work together to make the Church radiant and strong.

This is not just a study of church offices. It is an invitation to know Christ more deeply. As we explore the gifts, may we encounter the Giver. As we reflect on the roles, may we fall more in love with the One who fulfills them all perfectly. And as we grasp the beauty of God's design, may our hearts rise in awe.

Friend, prepare your heart. Let's step into this journey with humility, reverence, and expectancy. For the Five-Fold Ministry is not about us—it is about Him.

Prayer

"Oh Lord, I pray that our hearts may be encouraged, being knit together in love, and into all the wealth and riches of being entirely confident of comprehending the knowledge and mystery of God, both of the Father and of Christ, in whom are hidden all the treasures of wisdom and knowledge." ~Colossians 2:2,3 (TBV)

Chapter 1

The Apostle: Christ's Sent Ones

"According to the grace of God which was given to me, as a wise master builder I have laid the foundation, and another builds on it. But let each one take heed how he builds on it."
~1 Corinthians 3:10 (NKJV)

A Foundation Stone

Have you ever noticed that God never starts with the walls? He always begins with the foundation. You can see this in creation itself. Before light ever shone, before land appeared, before mankind took breath, the Spirit hovered over the waters. He was preparing the ground, establishing order, laying a foundation.

The apostolic gift carries this same heartbeat. Apostles are foundation-layers, entrusted with the weight of beginnings that will outlast themselves. The Greek word *apostolos* means "one who is sent." Yet it is more than being dispatched; it carries the sense of representing the sender, of bearing their authority. An apostle is one sent not in their own name, but in the name of the One who sent them.

And isn't that what Jesus did? Hebrews 3:1 calls Him the "Apostle and High Priest of our confession." He was the Sent One of the Father, carrying Heaven's authority to earth. Every apostle who has come since is simply reflecting that eternal reality—Jesus, the true Foundation.

Builders and Fathers

Think of Paul's letters. He wasn't just managing churches; he was laying blueprints. He called himself a wise master builder (1 Corinthians 3:10) and, at the same time, a spiritual father (1 Corinthians 4:15). That combination—builder and father—is at the core of apostolic ministry.

Builders create structure, order, and strength. Fathers nurture, protect, and raise up children into maturity.

Apostolic ministry carries both. Apostles design with eternity in mind, but they also walk closely with people in the here and now.

I know of a church plant that had just begun meeting in a storefront. The leaders were full of zeal, but the foundation was shaky. There was no discipleship process, no grounding in Scripture—just excitement. An apostolic leader came alongside, not to criticize, but to patiently guide them. He drew their eyes back to Christ as the cornerstone. He taught them to build on the Word, not just emotion. Years later, that church has multiplied into several healthy congregations. Why? Because a foundation was laid.

Beloved, that's what apostles do. They look at what is being built and say, "Is Christ the center? Are we aligned with His blueprint?"

Apostolic Authority: Strength Wrapped in Humility

Authority is a tricky word in our culture. Too many have seen it abused, and we often equate authority with control. But true apostolic authority looks like Jesus with a towel around His waist, washing feet. It is strength wrapped in humility, power expressed through service.

Paul had immense authority, but he described himself as a servant of Christ and even a prisoner of the gospel (Ephesians 3:1). His authority was not about domination; it was about responsibility. Apostles don't build empires for themselves. They labor for Christ's kingdom.

Have you ever been around someone who carried real spiritual authority? You probably felt two things at once: a holy weight that called you higher, and a safety that drew you near. That's apostolic. It challenges us to rise while anchoring us in Christ's love.

Apostles as Pioneers

Apostles are pioneers. They are wired to go where others haven't gone. Paul longed to preach Christ where He was not yet named (Romans 15:20). That same restlessness lives in apostolic people today.

I once knew a missionary couple who moved into a city where there were no thriving churches. Everyone told them it would never work. The soil was too hard, the people too resistant. But they carried an apostolic spirit. They prayed, labored, built relationships, and within a decade, not only was there a church, but there were multiple fellowships led by local believers. That's the mark of apostolic grace: they don't just blaze trails, they establish households of faith that last.

The apostle's vision is not just expansion, but sustainability. They don't want a fire that flares up and burns out—they want a fire that keeps burning for generations.

The Apostolic Mindset for Every Believer

Now you may be thinking, "But I'm not an apostle." That may be true in the sense of office, but every believer is called to live with an apostolic mindset.

- **Living Sent** – Do you know that you've been sent? Into your workplace, your family, your city? You carry Christ wherever you go.
- **Thinking Long-Term** – Apostles remind us to look beyond today's struggles. What choices am I making that will matter in eternity?
- **Building Carefully** – Am I constructing my life on Christ the solid rock, or on the shifting sands of culture and convenience?
- **Caring as Fathers and Mothers** – Who am I investing in? Whose faith am I nurturing so they too will stand strong?

Friend, you don't need a title to live sent. You only need obedience to the One who sends.

Consider This

- Am I building my life on the eternal foundation of Christ, or on temporary comforts?
- Do I live with a sense of mission—that my steps are ordered and purposeful?
- In what ways can I help lay a foundation in someone else's life?

Pause for a moment and let those questions rest on your heart. Don't rush past them. Sometimes the Spirit whispers in these quiet pauses more than in paragraphs of teaching.

Prayer

Lord Jesus, You are the Chief Apostle, the One sent from the Father to redeem us. Thank You for entrusting Your Church with apostolic leaders who lay foundations and call us to maturity. Teach me to live with an apostolic mindset—to see myself as sent, to build carefully on You as the Cornerstone, and to invest in others so they too will stand strong. May my life bear fruit that outlasts me, fruit that remains for Your glory. Amen.

Chapter 2

The Prophet: The Voice of God's Heart

"Surely the Lord God does nothing, unless He reveals His secret to His servants the prophets."
~Amos 3:7 (NKJV)

God Still Speaks

Have you ever noticed how much God longs to communicate with His people? From the first words, *"Let there be light,"* to the still, small whisper heard by Elijah (1 Kings 19:12), our God is a speaking God. He does not remain silent. He reveals His heart, His plans, and His ways. The prophetic gift is one of the clearest signs of this truth—God still speaks.

The Hebrew word for prophet, *nabi*, means "to bubble up" or "to flow forth." It carries the sense of an overflow, a message that cannot be contained. A prophet is not someone who invents words for God, but someone who receives the overflow of His heart and declares it faithfully.

At its core, prophecy is about intimacy. The prophet leans close enough to hear the heartbeat of Heaven, then turns and speaks what they have heard. That's why prophets often carry both burden and beauty—because they feel what God feels.

Prophets in Scripture

Think of Elijah calling fire from heaven, Isaiah painting visions of the coming Messiah, Jeremiah weeping with God's grief, or Agabus in Acts foretelling famine and Paul's imprisonment. The prophetic ministry has always been central in shaping God's people.

Notice how different these prophets were—some dramatic, some poetic, some tender, some fiery. The prophetic gift is not a single mold but a reflection of God's vast personality. What unites them is their role: they are voices that call the people back to God, guarding them from deception and realigning them with truth.

A Voice, Not Just a Prediction

Too often we reduce prophecy to foretelling the future, but the prophet's role is far more expansive. A prophet:

- Unveils God's perspective in confusing times.
- Encourages, strengthens, and comforts. (1 Corinthians 14:3)
- Guards against deception by reminding the Church of God's Word.
- Reveals God's heart to His people through the Word and personal prophecy.
- Unveils the secrets buried in people's hearts of His purposes for their lives.

Prophecy, when genuine, carries the fragrance of Christ. It never contradicts Scripture, it never glorifies self, and it always leads people closer to the Lord.

I remember hearing a simple prophetic word given during a small gathering. It wasn't flashy—it was just one woman saying, "The Lord wants you to know He sees your weariness and has not forgotten you." The person receiving it broke down in tears. Why? Because it was the right word at the right moment, carrying the weight of Heaven. That's what prophecy does. It pierces beyond human encouragement into divine confirmation.

The Cost of the Prophetic

The prophetic calling is not glamorous. Jeremiah was thrown into a pit. Elijah faced loneliness and despair. Even Jesus, the ultimate Prophet, was rejected in His hometown. Prophets often feel the ache of Heaven and wrestle with rejection from people.

But oh, what a gift they are to the body of Christ! Without the prophetic voice, the Church drifts into complacency, settling for the ordinary instead of pressing toward God's best. Prophets keep our eyes lifted, our ears attentive, our hearts tender.

The Prophetic Spirit for Every Believer

You may not stand in the office of a prophet, but Scripture tells us that we can all prophesy (1 Corinthians 14:31). What does this mean? It means that by the Spirit, every believer can speak words of encouragement, truth, and revelation that align with God's heart.

- **Listening Hearts** – cultivating quietness to hear His whisper.
- **Scripture Anchored** – testing everything against the Word so we remain grounded.
- **Encouraging Others** – speaking life when the Spirit nudges, even in simple ways.
- **Discerning Times** – asking, "Lord, what are You saying right now?"

Imagine if the entire Church leaned into this. Not everyone as prophets in office, but everyone carrying a prophetic spirit of encouragement and discernment. What a radiant, responsive bride we would be!

Guardrails for Prophecy

Because the prophetic gift is powerful, it must also be stewarded carefully. Paul told the Thessalonians, *"Do not despise prophecies. Test all things; hold fast what is good."* (1 Thessalonians 5:20–21).

Prophecy must be tested—against Scripture, by the witness of the Spirit, and in the counsel of mature believers. True prophets are humble enough to be corrected, patient enough to wait, and secure enough to let God vindicate their words in His time.

Consider This

- Am I cultivating space to hear God's voice regularly, or is my life too noisy?
- When I hear His whisper, do I obey and share, or do I shrink back in fear of rejection?
- Do I encourage others by speaking life, truth, and Scripture over them?
- How can I grow in discernment to recognize what is truly from the Lord?

Prayer

Lord, thank You that You are not silent. You speak, You reveal, You guide. Thank You for the gift of prophets who call us back to Your heart. Give me boldness to speak encouragement when You prompt me and humility to remain anchored in Your Word. May I carry a prophetic spirit that lifts others into Your presence. Amen.

"Hear me, O Judah and you inhabitants of Jerusalem: Believe in the Lord your God, and you shall be established; believe His prophets, and you shall prosper.
~2 Chronicles 20:20b (NKJV)

Chapter 3

The Evangelist: Bearer of Good News

"But you be watchful in all things, endure afflictions, do the work of an evangelist, fulfill your ministry."
~2 Timothy 4:5 (NKJV)

Carriers of Joyful News

Have you ever noticed how people light up when they hear truly good news? Whether it's the birth of a child, the healing of a sickness, or the homecoming of a loved one—good news changes the atmosphere. This is the heart of the evangelist: to carry and declare the greatest news ever told—that Jesus Christ saves, heals, and redeems.

The Greek word for evangelist, *euangelistēs*, means "a bringer of glad tidings." An evangelist is someone who overflows with the message of salvation, someone who can't help but share the love of Jesus with those who haven't yet encountered Him.

Evangelists are often described as the "harvesters" of the Five-Fold Ministry. Where apostles lay foundations, prophets call to alignment, pastors nurture, and teachers ground, the evangelist goes out into the fields with sickle in hand, calling the lost into the embrace of Christ.

Biblical Portraits of Evangelists

In the book of Acts, we see Philip called an evangelist (Acts 21:8). He was the one who preached in Samaria with such power that the entire city was filled with joy (Acts 8:8). He was also the one God sent to the desert road, where a single Ethiopian eunuch was waiting to hear the gospel. Isn't that beautiful? The evangelist is just as willing to preach to a crowd as they are to stop for the one.

Timothy, too, was charged by Paul to "do the work of an evangelist" (2 Timothy 4:5). Even though his role was pastoral, Paul reminded him that every shepherd must carry an evangelistic heart.

Above all, Jesus Himself embodied the evangelist's spirit. He proclaimed, *"The Spirit of the Lord is upon me, because He has anointed me to preach good news to the poor"* (Luke 4:18, NKJV).

Burning Hearts for the Lost

The evangelist carries a unique burden: a fire in their bones to see the lost saved. While others may grow comfortable inside church walls, evangelists are restless for those outside. They can't sit still knowing there are still people who haven't encountered the name of Jesus.

I remember a friend who could strike up a conversation with anyone—a cashier, an Uber driver, a stranger on the street—and somehow, within minutes, she would be talking about Jesus. Not because she was forcing it, but because her heart overflowed with it. That's what evangelists do. They make Jesus known, naturally and joyfully.

Evangelism in Everyday Life

You may say, "But I'm not an evangelist." That may be true in terms of office, but every believer is called to bear witness to Christ. Evangelism is not just for the fiery preacher on the street corner—it's for the quiet neighbor who invites a friend to church, the coworker who shares hope over coffee, the student who lives out their faith at school.

Imagine the ripple effect if every believer carried a simple readiness to share Jesus. Not pressured, not scripted, but Spirit-led and genuine. That's the evangelistic spirit at work in the whole body.

The Evangelist's Role in the Church

Evangelists are more than soul-winners—they are equippers. Ephesians 4:12 reminds us that all five gifts exist *"for the perfecting of the saints, to the work of serving."* Evangelists stir the whole Church to carry the gospel. They remind us that the harvest is ready, that eternity is real, and that love must compel us outward.

Without evangelists, the Church grows inward, consumed with itself. With evangelists, the Church grows outward, alive with mission. We need their fire. We need their urgency. We need their reminder that the gospel is still the power of God unto salvation (Romans 1:16).

The Eternal Perspective

Do you know that heaven throws a party every time one sinner repents (Luke 15:10)? Evangelists live with this eternal joy. They know that every soul is priceless, that eternity hangs in the balance. Their urgency is not born of fear, but of love.

One day, evangelism as we know it will end—because all will have seen the King in His glory. But until that day, the evangelist's cry keeps the Church awake: "Lift up your eyes! The fields are white for harvest!" (John 4:35, TBN).

The Evangelistic Spirit for Every Believer

Here are some simple ways we can all live evangelistically:

- **Pray for the Lost** – Begin with one or two names. Consistently lift them before the Lord.
- **Live Authentically** – Let your life speak before your words do. People are drawn to Christ in you.
- **Be Ready** – Peter said, *"Always be ready to give an answer to anyone who asks you a reason concerning the hope that is in you"* (1 Peter 3:15, WEB).
- **Share the Story** – Sometimes all it takes is, "Here's what Jesus did in my life." Your testimony carries power.

Consider This

- Do I carry God's heart for the lost, or have I grown comfortable inside the walls of the church?
- Who in my life might be waiting to hear a word of hope from me?
- How can I support and learn from those with an evangelistic calling?
- What simple steps can I take today to share Christ's love in word and deed?

Prayer

Lord Jesus, You are the Good News—the gospel made flesh. Thank You for the gift of evangelists who burn with passion to see the lost found. Stir my heart with Your compassion. Give me eyes to see the harvest fields around me and courage to share Your love. May my life be a testimony of Your grace, and may I never forget the joy of heaven when even one sinner repents. Amen.

Chapter 4

The Pastor: Shepherd of Souls

"I am the good shepherd. The good shepherd gives His life for the sheep."
~John 10:11 (NKJV)

The Heart of the Shepherd

Have you ever noticed how comforting it is when someone truly sees you—not just your successes or failures, but your heart? That is the essence of pastoral ministry. A pastor is one who tends, nurtures, and protects God's people, just as a shepherd watches over his flock.

The Greek word for pastor, *poimēn*, literally means "shepherd." It evokes the picture of someone guiding sheep to green pastures, guarding them from danger, and binding up the wounded. The pastoral gift is Christ's own compassion extended through His people. Jesus declared, *"I am the good shepherd"* (John 10:11). Every pastor is meant to mirror His shepherding love.

Shepherds in Scripture

David, before he was ever a king, was a shepherd boy who cared for his father's flock. He risked his life against lions and bears to protect the sheep (1 Samuel 17:34–35). That same protective, sacrificial heart would later mark his reign over Israel.

In the New Testament, Peter exhorted elders, *"Shepherd the flock of God which is among you, exercising the oversight... not for dishonest gain, but willingly;"* (1 Peter 5:2, WEB). Pastors are not to dominate, but to serve willingly, guiding with humility and tenderness and not for personal gain.

Isn't it beautiful how God so often entrusts leadership to those who first learn to care for sheep? The skills of guiding, protecting, and nurturing are not glamorous, but they are the very heart of His kingdom.

The Care of the Pastor

Pastors carry a burden for souls. They feel the ache of people's wounds, the weight of their struggles, the joy of their victories. While an apostle might see the big picture, and a prophet might thunder a call to repentance, the pastor bends low, tending to individual hearts.

I recall visiting a church where the pastor knew everyone by name—and not just their names, but their stories. He remembered which family was walking through grief, which teenager was struggling in school, which couple was trying to reconcile. He didn't just preach sermons; he carried their lives in prayer and presence. That is pastoral love.

Pastoral ministry is not flashy. It rarely makes headlines. But it is indispensable. Without shepherds, the flock scatters. With them, the people of God find safety, nurture, and rest.

The Cost of Shepherding

Shepherding is not easy. Sheep wander. They get themselves tangled in thorns. They sometimes bite the very hand that feeds them. Yet the pastor does not abandon them. Just as Jesus leaves the ninety-nine to go after the one (Luke 15:4), true pastors pursue, rescue, and restore—even when it costs them.

This is why Jesus said the good shepherd lays down his life for the sheep (John 10:11). Pastoral love is sacrificial. It is patient. It stays when others walk away.

The Pastoral Spirit for Every Believer

Not all are called to the office of pastor, but every believer can carry a pastoral spirit. Paul urged the Galatians, *"Bear one another's burdens, and so fulfill the law of Christ"* (Galatians 6:2, NKJV). We are all called to care for one another, to walk together in love.

- **Listening Ears** – taking time to hear people's hearts without rushing.
- **Protective Love** – standing up for the vulnerable, shielding them from harm.
- **Healing Presence** – being there in seasons of loss, grief, or confusion.
- **Community Builders** – fostering fellowship so no one walks alone.

Imagine a Church where every member lived with this shepherding spirit—caring, protecting, encouraging. What a reflection of Christ that would be!

The Pastor's Role in the Five-Fold

The pastoral gift holds the body together in love. Apostles build, prophets guard, evangelists reach out, teachers ground—but pastors knit the flock into family. They ensure that no one is left behind, that discipleship is not just about information but about transformation through relationship.

Without pastors, churches risk becoming organizations instead of families. With them, the Church feels like home.

Consider This

- Do I allow myself to be shepherded, or do I resist care and accountability?
- How can I cultivate a pastoral spirit toward others—bearing burdens, showing compassion and offering encouragement?
- Who around me may be wandering, weary, or wounded, waiting for someone to notice?
- Am I willing to lay down my comfort to walk alongside someone in their need?

Prayer

Lord Jesus, You are the Good Shepherd who knows me by name. Thank You for the gift of pastors who mirror Your care, who walk with us in our joys and sorrows. Teach me to receive their nurture with humility and to extend that same shepherding love to others. Help me to carry a pastoral spirit in my daily life—listening, protecting, encouraging, and building community. May Your flock be strengthened, and may You be glorified as our Chief Shepherd. Amen.

Chapter 5

The Teacher: Grounding in Truth

"And he continued there a year and six months, teaching the word of God among them."
~Acts 18:11 (WEB)

Anchoring in the Word

Have you ever noticed how quickly we forget things? A verse may stir us one day, but by the next, the worries of life crowd it out. This is why we need teachers. Teachers anchor us in God's Word, patiently laying truth upon truth until it sinks deep into our hearts.

The Greek word for teacher, *didaskalos*, means "instructor" or "one who imparts knowledge." But in the biblical sense, it is so much more than relaying facts. A teacher unfolds the Scriptures, opening eyes to see what was hidden. They don't just pass on information—they impart revelation, so that truth takes root and transforms lives.

Jesus Himself was called Teacher more than any other title. Everywhere He went, He was teaching—on mountainsides, in synagogues, in boats, on roadsides. His teaching was not dry or abstract; it was alive, practical, and Spirit-breathed.

Teachers in Scripture

Think of Ezra, who devoted himself to study, to live out, and to teach the law of God (Ezra 7:10). Or Apollos in Acts 18, described as "mighty in the Scriptures" and teaching with zeal, though later corrected and sharpened by Priscilla and Aquila. Teachers play a vital role in keeping the Church grounded in sound doctrine and aligned with God's Word.

Paul himself was both apostle and teacher (1 Timothy 2:7). He reminded Timothy to *"preach the word... reprove, rebuke, and exhort, with all patience and teaching"* (2 Timothy 4:2). Teaching requires patience—it is line upon line, precept upon precept, until faith is firmly established.

The Role of Teachers Today

Teachers bring clarity. They untangle confusion, explain context, and connect the dots so that Scripture is not only read but understood. In a world filled with voices, opinions, and distortions, teachers remind us of what God actually said.

I remember sitting under a teacher of the Word who explained a single verse for over an hour, but it felt like five minutes. The way he unpacked the original language, connected it with Old Testament foreshadows, and applied it to everyday life—suddenly, a passage I had skimmed for years came alive. That's the gift of the teacher. They don't just explain—they illuminate.

Teachers also protect the flock from error. When strange doctrines creep in, when the culture twists truth, teachers rise up with the Word as a plumb line. They are guardians of sound doctrine, ensuring that the Church is not swayed by every wind of teaching (Ephesians 4:14).

Teaching as Transformation

But here's the key: true teaching is not just head knowledge—it is heart transformation. It's possible to know facts about Scripture and yet not know the God of Scripture. Teachers remind us that the Word is living and active (Hebrews 4:12). They invite us not just to study but to encounter.

Jesus taught with authority, not like the scribes (Mark 1:22). His teaching carried weight because it was Spirit-breathed. Teachers today are called to carry that same Spirit-led authority, teaching not from pride of knowledge but from a life rooted in Christ.

The Teacher's Spirit for Every Believer

Not everyone is called to the office of teacher, but every believer is called to be a student and sharer of truth. We are all invited to learn the Scriptures, to apply them, and to pass them on.

- **Students of the Word** – Do I carve out time to study Scripture, not just skim?

- **Living the Word** – Do I practice what I know, or only talk about it?
- **Sharing the Word** – Do I explain truth to others, even simply, so that they too can understand?
- **Guarding Truth** – Do I measure teachings and voices against the plumb line of Scripture?

The Church flourishes when every believer takes responsibility for knowing and sharing the Word. Teachers ignite this hunger in us, but we must steward it personally.

The Eternal Perspective

One day, we will know fully, even as we are fully known (1 Corinthians 13:12). In eternity, teaching as we know it will cease—because we will see the Living Word face to face. But until that day, teachers are a grace to keep us steady, growing, and rooted in Christ.

Teachers remind us that truth is not relative, but eternal. They help us to see that the Word of God is not just ancient history, but present reality and eternal promise.

Consider This

- Am I growing in knowledge of God's Word, or am I relying on others to feed me?
- Do I see teaching as information, or as invitation to transformation?
- How can I cultivate a hunger for Scripture that goes beyond duty into delight?

Prayer

Lord Jesus, You are the Living Word, and You taught with authority and grace. Thank You for the gift of teachers who unfold the Scriptures and anchor us in truth. Stir in me a hunger for Your Word. Help me not only to hear, but to live what You teach. Guard my heart from deception and use me to share Your truth with others in love. May Your Word dwell richly in me and bear fruit that lasts into eternity. Amen.

Chapter 6

Unity of the Five: One Body, One Spirit

"He makes the whole body fit together perfectly. As each part does its work, it helps the other parts grow, so that the whole body is healthy and growing and full of love."
~Ephesians 4:16 (WEB)

Christ the Head, the Five as His Hand

Have you ever tried to grab something without using all your fingers? It feels awkward, incomplete. That is what the Church is like when one or more of the Five-Fold gifts is missing. Each gift is like a finger—distinct, purposeful, necessary—but all connected to the same hand, guided by the same Head, Christ Jesus.

The apostle pioneers and lays foundations. The prophet calls us back to God's heart. The evangelist carries good news to the lost. The pastor nurtures and protects. The teacher grounds us in truth. Five different expressions, but one Spirit, one Lord, one purpose: to reveal the fullness of Christ and prepare His Bride.

Balance, Not Competition

In too many places, the gifts have been seen as competitors rather than companions. Some elevate the prophet while dismissing the teacher. Others exalt the evangelist while overlooking the pastor. But this was never God's design.

The apostle without the pastor risks becoming visionary but cold.

The prophet without the teacher may thunder truth without grounding.

The evangelist without the apostle may gather crowds without structure.

The pastor without the prophet may comfort but never challenge.

The teacher without the evangelist may inform but not inspire.

When they function together, balance is restored. Like a symphony, each instrument plays its part. Alone, the notes may sound pleasant. Together, they create music that moves the soul.

The Danger of Imbalance

History is filled with examples of imbalance. Some movements were rich in teaching but lacked evangelistic fire, so they became stagnant. Others burned with prophetic zeal but collapsed because no apostolic foundation was laid. Still others drew massive crowds through evangelism, but without pastors and teachers, those new believers drifted away.

I remember visiting a church that leaned heavily on prophetic ministry. Every service was filled with fiery words, but people were spiritually weary because no one was grounding them in teaching or nurturing them pastorally. Later, I visited another congregation where teaching was strong, but evangelism was almost non-existent. They knew the Word well but had lost the urgency for the lost. Both churches loved the Lord, but both revealed the danger of imbalance.

Friend, we need the whole hand, not just a few fingers.

The Eternal Goal of Unity

It's easy to forget that the Five-Fold gifts are not the goal. They are means to the goal: the maturity of the Church. Paul wrote that these gifts were given "*until we all attain to the unity of the faith... to the measure of the stature of the fullness of Christ*" (Ephesians 4:13, NKJV).

This is about Christ, not us. The gifts are not badges of honor or positions of prestige. They are expressions of Jesus Himself, distributed among His people so that no one part of the body is lacking.

One day, when we see Him face to face, the gifts will no longer be needed. Prophets will be silent, because the Word Himself will stand before us. Evangelists will rest, because every soul will know Him. Teachers will bow, because perfect knowledge will flood us. Pastors will rejoice, because the flock will be gathered forever. Apostles will lay down their tools, because the city of God will be complete. But until that day, we need all five—working together, pointing us to Him.

Practical Application for Believers

So what does this mean for us, right now, in our daily walk with Christ?

- **Honor the Gifts** – Don't dismiss the ones you find unfamiliar. If you naturally love teachers, learn to value prophets. If you lean toward pastors, don't neglect evangelists. All are from Christ.
- **Submit to the Diversity** – Sometimes the gift that stretches us the most is the one we need most. Be willing to receive even when it challenges your comfort.

- **Celebrate Unity** – Rejoice in the variety of Christ's expression in His body. The diversity of gifts is a testimony to His fullness.
- **Discern Your Shaping** – Ask: Which of these gifts has God used most to shape me? Where might I need balance from the others?

Imagine a Church where every believer embraced the fullness of Christ's gifts—receiving, honoring, and learning from each one. What maturity! What beauty!

Consider This

- Do I tend to elevate one of the Five-Fold gifts while neglecting the others?
- Where have I seen imbalance in the Church, and how did it affect growth?
- How can I personally help foster unity—honoring all the gifts and pointing people back to Christ?
- Am I open to being shaped by all five expressions of His grace?

Pause and let these questions settle in your heart. Unity is not just a concept—it's a call to humility, honor, and surrender.

Prayer

Lord Jesus, You are the Head of the Church, and we are Your body. Thank You for giving apostles, prophets, evangelists, pastors, and teachers to equip and mature us. Forgive us when we elevate one gift above the others or despise what we do not understand. Teach us to honor every expression of Your grace, to walk in unity, and to fix our eyes on You above all. Prepare Your Bride, Lord, to be healthy, growing, and full of love until the day we see You face to face. Amen.

Conclusion

The Bride Made Ready

"Let us be glad and rejoice and give Him glory, for the marriage of the Lamb has come, and His wife has made herself ready."
~Revelation 19:7 (NKJV)

The Gifts and the Giver

As we've walked through the five gifts of Christ—the apostle, prophet, evangelist, pastor, and teacher—we've seen how each one carries a part of His heart. The apostle builds. The prophet listens. The evangelist proclaims. The pastor nurtures. The teacher grounds.

But here is the greater truth: the Five-Fold Ministry is not ultimately about people or titles. It is about Jesus. He is the Apostle sent from the Father, the Prophet who reveals God's heart, the Evangelist who brings good news, the Pastor who lays down His life for the sheep, and the Teacher who is Himself the Word made flesh.

When we honor these gifts, we honor Him. When we receive from them, we receive Him. And when they work together in unity, the Church begins to look more and more like Christ in the world.

From Earth to Eternity

Paul reminds us that these gifts were given *"until we all attain to the unity of the faith... to the measure of the stature of the fullness of Christ"* (Ephesians 4:13, WEB). That little word *until* matters. It points us forward, beyond this age, into eternity.

For now, apostles still labor to lay foundations. Prophets still call us back to God's heart. Evangelists still cry out for the lost. Pastors still tend weary souls. Teachers still unfold the Word. But one day—oh, what a day it will be!—their work will be complete.

In eternity, we won't need foundations, for the city of God will stand forever. We won't need prophecy, for we will see Him face to face. We won't need evangelism, for all will know Him. We won't need pastors, for the Great Shepherd will hold us. We won't need teachers, for perfect knowledge will flood us like light.

Until then, we need these gifts. Until then, we embrace them with gratitude. Until then, we let them shape us into the Bride made ready.

Our Call as Believers

What does this mean for you and me? It means we live with an open heart, willing to be shaped by all of Christ's expressions.

- We let apostles stretch our vision beyond ourselves.
- We let prophets call us back when we drift.
- We let evangelists stir our passion for souls.
- We let pastors walk with us in weakness and joy.
- We let teachers ground us in eternal truth.

And we, in turn, take responsibility to live sent, to listen, to proclaim, to nurture, and to teach in the measure of grace God has given us.

Friend, you may not carry a formal office, but you carry His Spirit. And His Spirit is more than enough to use you to build, to encourage, to proclaim, to shepherd, and to teach someone else.

The Eternal Perspective

When John saw the vision of heaven in Revelation 19, he saw a Bride made ready. That is the end goal of the Five-Fold Ministry—not busier churches, not bigger platforms, not better programs, but a people prepared for their Bridegroom.

Every gift, every message, every labor, every tear poured out by these servants of Christ is aimed at that one day—the day when the Bride stands radiant, clothed in fine linen, pure and bright, ready for the wedding of the Lamb.

Oh, beloved, let us live with that day in view. Let us not despise the gifts, nor idolize them, but receive them as God's provision to make us mature, holy, and ready for eternity.

Consider This

- Am I living with eternity in view, or only with today's concerns?
- How has God used the Five-Fold gifts to shape me? Which ones do I need to open my heart to more fully?
- Am I willing to honor and support all the gifts, even those that stretch or challenge me?
- How can I take what I have learned and live it out so that others are strengthened in Christ?

Closing Prayer

Lord Jesus, You are the fullness of every gift. You are our Apostle, our Prophet, our Evangelist, our Pastor, and our Teacher. Thank You for giving these expressions of Your heart to Your Church. Thank You for leaders who labor, listen, proclaim, nurture, and teach in Your name.

Make me humble enough to receive from all that You give. Shape me through their voices, their service, their example. Let me carry a portion of Your heart to others, so that they too may grow in You.

And Lord, keep my eyes on eternity. Prepare me, prepare us, prepare Your Bride, until the day we stand before You, radiant and ready, rejoicing in the wedding of the Lamb. Amen.

Sharing Jesus & the Holy Spirit with Others

Jesus gave us the mandate to go and share the good news and to make disciples with this newfound authority that He has given us. In Matthew 28:18-20, "Jesus came to them and spoke to them, saying, "All authority has been given to me in heaven and on earth. 19Go and make disciples of all nations, baptizing them in the name of the Father and of the Son and of the Holy Spirit, 20teaching them to observe all things that I commanded you. Behold, I am with you always, even to the end of the age." Amen."

Here is a simple prayer you can share: "Jesus, I believe that You died for me and that You rose again after three days. I confess to You that I am a sinner and in need of a Savior. I need your love, forgiveness and cleansing to make me whole. Come into my life, forgive my sins and give me a life that will never end with You. Come and take control of my life and make me the person You created me to be. I believe that You are preparing a place for me in heaven and that You will be returning soon. I confess You now as my Lord and Savior. Thank you for saving me! Amen."

And to introduce them to the Holy Spirit, lead them in this prayer: "Father, thank you for sending Your Son, Jesus, to give me eternal life. Now I ask You to fill me with Your Holy Spirit to empower me and to teach me Your ways. Immerse me and fill me to overflowing with rivers of living water that will flow from my innermost being speaking in a new heavenly language. I receive this beautiful gift that You desire to give me. Thank you! In Jesus' name I pray. Amen."

We have the guarantee in God's Word that He desires to give us good gifts and the greatest gift He could give is His Son, Jesus Christ (John 3:16) and His Holy Spirit (Luke 11:13). He is the best Gift-giver and He will give the Holy Spirit to those who ask Him just as He gave Jesus freely to save us. It is by faith that you receive Jesus. It is by faith that you receive the Holy Spirit.

Ask Him. Receive the Holy Spirit. Your life will never be the same!

Your prayer life will go to a new place of faith and confidence in praying His perfect will when you don't know how or what to pray (Romans 8:26-28).

I trust that this book has been a blessing to you in your journey to know the value and purpose of the five-fold ministry as we are all being equipped for the work of the ministry.

With love and expectation for you to come into unity and knowledge of the fullness of Jesus Christ and His eternal purposes for your life.

~ Bonnie Sherwood

Made in the USA
Monee, IL
04 November 2026

67378669R00036